Falcon Heavy Launch of the X-37B Plane

A New Era of Military Spaceflight

Beatrice Fairchild

Table of Contents

Introduction

In the vast expanse of our universe, a monumental shift has begun—a paradigmatic leap propelling humanity into a realm where innovation and secrecy collide, birthing the dawn of a new era in military spaceflight.

Imagine a moment where the might of innovation and the veiled allure of classified technology unite, propelling humanity's quest beyond Earth's boundaries. The Falcon Heavy launch of the X-37B, a diminutive yet enigmatic spaceplane, stands at the forefront of this evolutionary journey—a venture shrouded in secrecy yet pulsating with promises of unparalleled technological prowess.

In this book, we embark on a compelling journey, exploring the nexus where cutting-edge space technology meets the clandestine world of military exploration. Each chapter unfurls the intricate tapestry of the Falcon Heavy rocket's design, the enigmatic features of the X-37B spaceplane, and the controversies that enshroud this clandestine program.

The relevance of this venture transcends the mere launching of a spaceplane; it encapsulates the zenith of technological innovation interwoven with the dilemmas of secrecy, cost, and the ethical boundaries of militarizing space. As an expert in space technology, my aim is to clarify the complexities, unveil the mysteries, and scrutinize the potential impacts of this pivotal launch.

Throughout these pages, we shall traverse the technicalities, decipher the controversies, and probe into the ethical considerations of the X-37B program. This book is not merely an exposé of a rocket's prowess; it's an invitation to the future—a glimpse into the uncharted domains of military spaceflight.

Buckle up as we delve into the Falcon Heavy launch of the X-37B, an enthralling expedition that promises to reshape our understanding of space technology, unveil secrets, and raise profound questions about the trajectory of our exploration beyond Earth's confines.

So, dear reader, fasten your seatbelts for a journey that transcends the boundaries of our atmosphere, where innovation meets intrigue, and secrecy dances with the cosmos. The mysteries unveiled within these pages shall unravel just a fraction of the enigmatic saga that awaits in the boundless expanse of space.

Join me as we embark on this awe-inspiring journey, a journey that opens the doors to the universe and challenges our perceptions of the final frontier.

Welcome aboard! 🪐

Chapter 1

The Falcon Heavy Rocket: Design and Capabilities

The Falcon Heavy rocket is a partially reusable super heavy-lift launch vehicle designed, manufactured, and launched by SpaceX. It is composed of three reusable Falcon 9 nine-engine cores whose 27 Merlin engines together generate more than 5 million pounds of thrust at liftoff, equal to approximately eighteen 747 aircraft. The rocket consists of a center core on which two Falcon 9 boosters are attached, and a second stage on top of the center core. The Falcon Heavy rocket is 70 meters (229.6 feet) tall and 12.2 meters (39.9 feet) wide, with a mass of 1,420,788 kg (3,125,735 lb).

Overview of the Falcon Heavy rocket's design and structure

The Falcon Heavy rocket, designed and developed by SpaceX, represents a significant advancement in space launch technology. Its design and structure showcase several innovative features aimed at achieving high performance, reusability, and safety standards.

Overall Design and Specifications:

Human Rating: The Falcon Heavy is engineered to meet or surpass current human-rating requirements, ensuring it meets stringent safety criteria for crewed missions.

Enhanced Safety Margins: Notably, the rocket boasts safety margins that are 40% above flight loads, which is significantly higher than the 25% margins found in other rockets. This emphasis on structural safety highlights SpaceX's commitment to reliability and safety.

Propellant and Engines: The first stage of the Falcon Heavy comprises three aluminum-lithium alloy rocket cores. These cores house a total of 27 Merlin engines that use a combination of liquid oxygen and rocket-grade kerosene (RP-1) as propellants. The use of multiple engines enhances both the rocket's thrust capabilities and redundancy in case of engine failures during ascent.

Landing Legs: Each of the three rocket cores on the first stage is equipped with four landing legs, totaling 12 legs. These cutting-edge landing legs are constructed using state-of-the-art carbon fiber combined with aluminum honeycomb structures. They are designed to withstand the extreme conditions of landing and facilitate the rocket's reusability by enabling precise landings on autonomous drone ships or landing pads.

Interstage: The interstage component serves as a crucial link connecting the center core of the first stage and the second stages. It houses the release and separation system, allowing for the clean separation of stages during the rocket's ascent.

Grid Fins: Falcon Heavy employs 12 hypersonic grid fins, with four fins attached to each booster. These grid fins, positioned at the base of the interstage or nosecone, play a pivotal role during reentry. They dynamically control and steer the rocket by altering the center of pressure, ensuring a controlled descent and facilitating precise landing maneuvers.

Engineering Innovations and Reusability:

The incorporation of multiple cores, numerous engines, advanced materials like aluminum-lithium alloys and carbon fiber, as well as innovative landing legs and grid fins, collectively contribute to the rocket's reusability and cost-effectiveness.

SpaceX's emphasis on reusability is a breakthrough in the space industry, aiming to reduce the expenses associated with space missions by recovering and reusing major components of the rocket.

Implications and Significance:

The Falcon Heavy's design and structure represent a leap forward in rocket technology, offering higher performance, improved safety margins, and reusability, potentially revolutionizing the space launch market.

Its ability to carry large payloads and its suitability for a wide range of missions, including commercial satellite launches, interplanetary missions, and potential crewed missions, make it a versatile and influential player in the realm of space exploration.

Capabilities of the Falcon Heavy rocket
Powerful Lifting Capacity:

Falcon Heavy is like a super strong crane in space. It's one of the mightiest rockets globally, capable of carrying nearly 64 metric tons (141,000 lbs) into orbit. Imagine it as a massive truck that can haul a lot of stuff into space at once.

Compared to its closest competitor, the Delta IV Heavy, Falcon Heavy can carry more than double the amount of cargo. It's like saying Falcon Heavy is way stronger than the next strongest rocket out there.

Versatile Travel Beyond Earth:

Not only can Falcon Heavy lift a lot of weight to Earth's orbit, but it can also go even farther into space. It's like saying it's not limited to dropping off things just in Earth's orbit - it can carry cargo to places beyond Earth too. This makes it handy for missions aiming to explore other planets or deep space.

Enormous Thrust and Multi-Stage Flight:

When Falcon Heavy blasts off, its first stage engines push out more than 5 million pounds of force, lifting it up into the sky. It's like having the power of several huge airplanes all at once.

After some time, when it's high enough, the first-stage parts separate, and the second-stage engine takes over. This second stage, called the Merlin Vacuum Engine, works in space where there's no air. It's like the engine that pushes your toy rocket after the big firework part has finished. This engine helps carry the cargo all the way to its destination in orbit around Earth or beyond.

So, in simple words, the Falcon Heavy is a giant rocket that can carry a lot more weight than most others, it can travel far beyond Earth, and it uses really powerful engines to lift off and reach its destination in space. It's like the superhero of rockets, capable of doing some really heavy lifting and going on incredible journeys through space!

Comparison with other rockets
Falcon Heavy's Superior Power:

Imagine Falcon Heavy as the strongest weightlifter among rockets. It's currently the most powerful rocket worldwide, and it's twice as strong as the next most powerful one, the Delta IV Heavy. Think of it as Falcon Heavy being able to lift a lot more weight in space compared to any other rocket right now. It's like one superhero being twice as strong as another superhero.

Carrying More Payload:

When we talk about "payload," it means the stuff the rocket can carry. Falcon Heavy is like a gigantic delivery truck that can carry more than double the amount of cargo than the Delta IV Heavy. It's like saying if Delta IV Heavy can carry two big boxes, Falcon Heavy can carry more than four of those same boxes. So, it's much better at hauling things into space.

Cost-Effectiveness:

Now, when we talk about the cost, Falcon Heavy is also much more wallet-friendly compared to the Delta IV Heavy. It's like saying if you wanted to buy two cool gadgets, one might cost you $350, but the other, which is just as good or even better, would only cost you $90. So, Falcon Heavy is a lot cheaper to launch into space compared to Delta IV Heavy, which makes it more affordable and cost-effective for missions.

The Falcon Heavy rocket is a super heavy-lift launch vehicle designed, manufactured, and launched by SpaceX. It is composed of three reusable Falcon 9 nine-engine cores whose 27 Merlin engines together generate more than 5 million pounds of thrust at liftoff, equal to approximately eighteen 747 aircraft. Falcon Heavy is capable of lifting nearly 64 metric tons (141,000 lbs) to orbit, more than twice the payload of the next closest operational vehicle, the Delta IV Heavy. Falcon Heavy is also more cost-effective than other heavy-lift rockets, with a launch cost of $90 million compared to the Delta IV Heavy's $350 million. The Falcon Heavy rocket is designed to meet or exceed all current requirements of human rating, and it is capable of carrying cargo beyond Earth orbit.

Chapter 2

The X-37B Spaceplane: Design and Capabilities

The X-37B spaceplane is an unmanned spacecraft designed and manufactured by Boeing for the United States Air Force. It is one of the world's newest and most advanced re-entry spacecraft, designed to operate in low-earth orbit, 150 to 500 miles above the Earth. The X-37B spaceplane is a reusable vehicle that explores reusable vehicle technologies that support long-term space objectives.

Overview of the X-37B's design and structure
Size and Shape:

The X-37B is a small spaceplane that looks a bit like the Space Shuttle but is only one-fourth its size. Imagine a smaller version of the Space Shuttle – like a mini spaceplane.

Combining Aircraft and Spacecraft Qualities:

The design of the X-37B combines the best parts of airplanes and spacecraft. It's made to be both affordable and easy to operate and maintain. It's like having a vehicle that's good at flying in space but is also simple to handle and doesn't cost a lot to run.

Innovative Technology for Space Operations:

This spaceplane uses a lot of new stuff for space travel. It has smart electronics that can control its descent and landing automatically without needing a lot of human input. Instead of hydraulic systems, it uses electronic controls and brakes, making it more modern and probably easier to keep working in space.

Also, the X-37B is made using different materials than usual. Instead of regular aluminum, it's built using lighter composite materials. It's like using new and lighter materials in a car to make it faster and more efficient.

Moreover, it has new kinds of tiles on its wings and body that can handle really high temperatures. These special tiles help protect the spaceplane during its travels through the hot re-entry into Earth's atmosphere.

Insulation and Protection:

To keep itself safe during flights, the X-37B has advanced insulating blankets and tough tiles that protect it from extreme heat. It's like having a cozy blanket to stay warm, but these are special ones to keep the spaceplane safe from the intense heat of re-entry into Earth's atmosphere.

In simple words, the X-37B is a smaller, smart spaceplane that combines the best of airplanes and spacecraft. It uses new technology, lighter materials, and special tiles to fly in space and return safely to Earth, making it efficient, easier to handle, and cost-effective for space missions.

Capabilities of the X-37B spaceplane
Long-duration Flights:

The X-37B is like a space traveler that can stay in space for quite a long time – up to 270 days! Imagine if a spaceship could go on a really long journey and stay out in space for nearly nine months.

Bringing Experiments Back to Earth:

Unlike many other spacecraft, the X-37B can carry experiments in space and then bring them back to Earth. It's like being able to take special things to space, test them out, and then safely bring them back to scientists on Earth for closer examination and study.

Special Solar Array for Long Missions:

This spaceplane has a special solar array that can be unfolded, like opening a big umbrella in space. This helps the X-37B stay in space for a really long time. It's like having a giant battery that keeps charging in the sunlight, allowing the spaceplane to fly for extended periods without needing to come back to Earth for refueling or recharging.

Safety Standards for Humans:

The X-37B is designed to meet or even exceed all the rules and safety standards required for human missions. Even though it's not carrying humans right now, it's built to be as safe as a spacecraft that would have people on board. It's like building a spaceship as safe as a car, even if there are no passengers inside.

Comparison with other spaceplanes
Advanced Re-entry Spacecraft:

The X-37B spaceplane is like one of the smartest space vehicles we have today for coming back to Earth. It's designed to work in a specific area of space, not too close and not too far – about 150 to 500 miles above the Earth. It's made so that it can bring experiments back from space for scientists to study more closely on Earth.

Cost-Effectiveness:

Compared to other spaceplanes like the Space Shuttle, the X-37B is much cheaper to launch. If launching the Space Shuttle cost about $1.5 billion, launching the X-37B only costs about $100 million. It's like comparing the cost of a super expensive car to a more reasonably priced one that works just as well.

The X-37B spaceplane is an unmanned spacecraft designed and manufactured by Boeing for the United States Air Force. It is one of the world's newest and most advanced re-entry spacecraft, designed to operate in low-earth orbit, 150 to 500 miles above the Earth. The X-37B spaceplane is a reusable vehicle that explores reusable vehicle technologies that support long-term space objectives. The X-37B spaceplane is capable of performing flights lasting up to 270 days and is equipped with a deployable solar array that allows extremely long flights. The vehicle is designed to meet or exceed all current requirements of human rating. The X-37B spaceplane is more cost-effective than other spaceplanes, with a launch cost of $100 million compared to the Space Shuttle's $1.5 billion.

Chapter 3

The Falcon Heavy Launch: Procedures and Challenges

The Falcon Heavy rocket is one of the most powerful operational rockets in the world, designed to lift heavy payloads into space. In this section, we will discuss the launch procedures for the Falcon Heavy rocket, the challenges faced during the Falcon Heavy launch, and compare it with other rocket launches.

Launch procedures for the Falcon Heavy rocket
Launch Location:

The Falcon Heavy rocket takes off from a place called the Kennedy Space Center in Florida, USA. It's like a big launching pad for rockets.

Launch Sequence:

When it's time for liftoff, the Falcon Heavy goes up into space just like the Falcon 9 rocket but with some changes. It's made up of three parts called cores, which are like big engines. Each of these cores has nine Merlin engines, and when they all work together, they make a huge force called thrust – more than 5 million pounds! It's like having the power of lots of big trucks pushing the rocket up into the sky.

The rocket is set up in a way that there's one big middle part called the center core, and two smaller ones are stuck on the sides. On top of the center core is another part called the second stage. It's like stacking three toy blocks on top of each other.

Control during Reentry:

As the Falcon Heavy comes back down to Earth, it has special fins called hypersonic grid fins. There are 12 of these, with four on each of the smaller side boosters. They're positioned at the bottom part of the rocket where the top meets the middle. These fins help control the rocket's direction by moving the center of pressure. It's like using a rudder on a boat to steer it in the right direction.

So, when the Falcon Heavy launches, it goes up from Kennedy Space Center just like the Falcon 9 rocket but with three big parts instead of one. These parts work together with lots of engines, and when it comes back down, it uses special fins to control where it lands.

Challenges faced during the Falcon Heavy launch
One big challenge for the Falcon Heavy rocket is making sure that the boosters, the parts that help the rocket go up, can come back and land safely on Earth. These boosters are made to land so that they can be used again for another launch. But it's not easy – it's like trying to land two big planes at the same time, and they have to land perfectly.

Getting the timing just right for them to land safely is really hard.

Another challenge is dealing with the weather. For the Falcon Heavy to launch properly, it needs the weather to be just right – clear skies and not too much wind. If the weather isn't good, like if it's too cloudy or too windy, it can cause a delay in the launch. Sometimes, they might even have to cancel the launch altogether, which is called a scrub.

So, the Falcon Heavy faces challenges like making sure the boosters can land safely after the launch, which is tough because it needs precise timing. Also, it needs perfect weather conditions for a successful launch, and if the weather isn't good, it can cause delays or even cancel the launch.

Comparison with other rocket launches
Powerful Lifting Capacity:

The Falcon Heavy rocket is like a super-strong giant among rockets. It's one of the mightiest rockets in the world, capable of lifting nearly 64 metric tons (that's like carrying the weight of many big elephants) into space. It's like saying this rocket is really good at hauling heavy stuff up to space compared to other rockets.

Cost-Effectiveness:

When it comes to money, the Falcon Heavy is much cheaper compared to other big rockets like the Delta IV Heavy. If the Delta IV Heavy costs around $350 million for a launch, the Falcon Heavy costs only $90 million. It's like saying if you had two things that did the same job, one would be way more expensive than the other. So, the Falcon Heavy is like the more affordable option for launching big things into space.

Reusable Boosters:

What's really cool about the Falcon Heavy is that it's designed in a way that after the launch, the boosters (those big parts that help the rocket go up) can come back down and land safely on Earth. This means they can be used again for another launch. It's like using the same car over and over instead of having to buy a new one each time. This feature of reusing boosters helps make the Falcon Heavy even more cost-effective because it saves a lot of money that would otherwise be spent on making new boosters for every launch.

The Falcon Heavy rocket is launched from the Kennedy Space Center in Florida, USA. The launch procedure for the Falcon Heavy rocket is similar to that of the Falcon 9 rocket, with some modifications. The Falcon Heavy rocket has faced several challenges during its launch history, including the successful landing of the rocket's boosters and weather conditions. The Falcon Heavy rocket is one of the most powerful operational rockets in the world today, capable of lifting nearly 64 metric tons (141,000 lbs) to orbit. The Falcon Heavy rocket is also more cost-effective than other heavy-lift rockets, with a launch cost of $90 million compared to the Delta IV Heavy's $350 million. The Falcon Heavy rocket's boosters are designed to land back on Earth after launch, making them reusable, which reduces the cost of manufacturing new boosters for each launch.

Chapter 4

The X-37B Spaceplane: Missions and Payloads.

The X-37B spaceplane is an unmanned spacecraft designed and manufactured by Boeing for the United States Air Force. It is one of the world's newest and most advanced re-entry spacecraft, designed to operate in low-earth orbit, 150 to 500 miles above the Earth.

Overview of the X-37B's missions

The X-37B spaceplane is like a versatile tool that's been used for lots of different jobs in space. It has done things like showing off new technologies, doing science experiments, and even helping out with military tasks. It's like having a robot that can do many different tasks, from cleaning to cooking.

This spaceplane is made to work in a specific area of space, not too close and not too far from Earth – about 150 to 500 miles above us. It's like having a flying car that drives on a specific road in the sky. This is where it does its job and comes back from to bring experiments for scientists to study.

To stay up in space for a long time, the X-37B uses a special solar array that unfolds like a big blanket. This helps the spaceplane stay up there for really long flights. It's like having a super long-lasting battery that charges in the sunlight, letting the spaceplane fly for a really long time without needing to come back down to Earth for more power.

Even though the X-37B doesn't carry people, it's built to be as safe as a spacecraft that would have people on board. It meets really high safety rules, so it's like building a very safe car even if there are no passengers inside.

So, the X-37B is like a handy space tool that does various jobs in space, from showing new technology to doing science experiments and military tasks. It operates in a specific area of space, stays up there for a long time with solar power, and is built to be very safe, making it a versatile and advanced spacecraft for different types of space missions.

Payloads carried by the X-37B
The X-37B spaceplane is like a delivery truck that takes lots of different things to space. It has carried all sorts of stuff on its missions, like special experiments and equipment.

Examples of Payloads Carried:

One thing it carried was an experiment called the Photovoltaic Radio-frequency Antenna Module (PRAM). This experiment changed sunlight into a special kind of energy called radio-frequency microwave energy. It's like using sunshine to make a different type of power.

There were also two other experiments from NASA. They looked at how things like materials and seeds are affected by being in space. It's like sending things to space to see how they change when they're away from Earth.

Additionally, it carried something called FalconSat-8, which is like a small satellite. This was developed by the U.S. Air Force Academy and sponsored by the Air Force Research Laboratory. It's like a tiny spaceship that does special tasks in space.

Versatility in Payloads:

The X-37B can carry lots of different things, like experiments, sensors (which are like machines that gather information), and other types of equipment. It's like being able to take all sorts of different tools and gadgets to space for different jobs.

So, in simple terms, the X-37B has taken lots of different things to space on its missions, from experiments changing sunlight into energy to studying how materials behave in space, and even small satellites. It's like a multi-purpose delivery van that carries various tools and experiments for different tasks in space.

Comparison with other spaceplanes

The X-37B is like one of the smartest spaceships for coming back to Earth that we have today. It's designed to work in a specific area of space, not too close and not too far from Earth – about 150 to 500 miles above us. It's made to bring experiments back from space so that scientists can study them more on Earth.

When it comes to money, the X-37B is much cheaper compared to other spaceplanes, especially like the Space Shuttle. If the Space Shuttle used to cost around $1.5 billion for one launch, the X-37B only costs about $100 million. It's like comparing the cost of a super expensive mansion to a nice house that costs much less. So, the X-37B is like the more affordable option for doing things in space.

Chapter 5

The Falcon Heavy Launch: Implications for Military Spaceflight

In this chapter, we'll discuss the significance of the Falcon Heavy launch of the X-37B, its implications for the future of military spaceflight, and compare it with other rocket launches.

Significance of the Falcon Heavy launch of the X-37B.

The Falcon Heavy launch carrying the X-37B is a big deal because it's the first time that the X-37B spaceplane has ever flown using the Falcon Heavy rocket with its three boosters. It's like trying something new, such as driving a different kind of car for the first time.

Powerful Rocket Capability:

The Falcon Heavy rocket is really powerful! It's like one of the strongest rockets in the world. It can carry nearly 64 metric tons into space. It's like saying it's a super strong truck that can lift very heavy things into space compared to other rockets.

Reusable Boosters for Cost Savings:

What's super cool about the Falcon Heavy is that after it launches, its boosters (the parts that help it go up) can come back and land safely on Earth. They're made to be used again for another launch. This is like using the same toy over and over instead of needing a new one each time. This helps save a lot of money because making new boosters for every launch can be really expensive.

So, the Falcon Heavy launch carrying the X-37B is special because it's a new way for the X-37B to go to space, using a really powerful rocket. Also, this rocket can bring its boosters back for reuse, which is great because it helps save a lot of money for future launches. Overall, it's a significant event for space exploration and cost-effective space travel.

Comparison with other rocket launches

The Falcon Heavy rocket is like a superhero among rockets. It's one of the strongest rockets in the world. It can carry nearly 64 metric tons into space. It's like a giant truck that can lift really heavy things into space, more than many other rockets can.

Compared to other big rockets like the Delta IV Heavy, the Falcon Heavy is much cheaper. If launching the Delta IV Heavy costs about $350 million, launching the Falcon Heavy only costs $90 million. It's like comparing the cost of a super fancy yacht to a more affordable boat. So, the Falcon Heavy is like the more affordable option for launching heavy stuff into space.

What's really smart about the Falcon Heavy is that after it launches, the boosters (those big parts that help the rocket go up) can come back and land safely on Earth. They're made so they can be used again for another launch. It's like using the same toy or tool over and over again instead of buying a new one each time. This saves a lot of money because making new boosters for every launch can be really expensive.

The Falcon Heavy launch of the X-37B is significant for several reasons. It marks the first time that the X-37B spaceplane has flown aboard the triple-booster Falcon Heavy rocket. The Falcon Heavy rocket is one of the most powerful operational rockets in the world, capable of lifting nearly 64 metric tons (141,000 lbs) to orbit. The Falcon Heavy rocket's boosters are designed to land back on Earth after launch, making them reusable, which reduces the cost of manufacturing new boosters for each launch. The Falcon Heavy launch of the X-37B has several implications for the future of military spaceflight, including the importance of reusable launch vehicles and the United States' continued commitment to developing advanced space technologies. The Falcon Heavy rocket is also more cost-effective than other heavy-lift rockets, with a launch cost of $90 million compared to the Delta IV Heavy's $350 million.

Chapter 6

The X-37B Spaceplane: Orbital Operations

Overview of the X-37B's orbital operations

The X-37B spaceplane is like a special vehicle designed to work in a specific area of space called low-earth orbit. This area is not too far and not too close to Earth – about 150 to 500 miles above us. It's like a high-altitude road in the sky where the spaceplane does its work and then comes back down to Earth.

This spaceplane is really smart because it can carry experiments up into space and then bring them back to Earth. Scientists can then study these experiments more closely here on our planet. It's like taking something on a trip and then bringing it back home for further examination.

To stay up in space for a very long time, the X-37B uses a special solar array that can be unfolded like a big blanket. This helps the spaceplane stay up there for really long flights. It's like having a super long-lasting battery that charges in the sunlight, letting the spaceplane fly for a really long time without needing to come back down to Earth for more power.

Even though the X-37B doesn't carry people, it's built to be as safe as a spacecraft that would have people on board. It meets really high safety rules, so it's like building a very safe airplane, even if there are no passengers inside.

Maneuvering capabilities of the X-37B

The X-37B spaceplane can do many different moves while orbiting Earth. It can change where it goes around Earth (called its orbit), move higher or lower (change its altitude), and do other moves as needed. It's like a very flexible and versatile dancer that can change its steps during a performance.

To make these moves, the spaceplane has something like a space engine that helps it move around. This engine allows the X-37B to change direction or speed up, just like how a car's engine helps it move on the road.

The X-37B also has a clever system that helps it know where it is and where it's going in space. It's like having a GPS system for space. This system helps the spaceplane figure out how to move and where to go while it's orbiting Earth.

Chapter 7

The Falcon Heavy Launch: Technological Advancements

Technological advancements in the Falcon Heavy launch of the X-37B

In this chapter, we'll discuss the technological advancements in the Falcon Heavy launch of the X-37B and its potential applications.

Lifting Heavier Payloads

The Falcon Heavy launch carrying the X-37B marked a big step forward in rocket power. It showed that this rocket can carry very heavy stuff into space. It's like a really strong truck proving it can lift much heavier things than before.

Importance of Reusable Rockets

This launch also taught us that using rockets that can be used again (like the Falcon Heavy's reusable boosters) is super important, especially for military space missions. Reusing rockets means it costs less money to send things into space. It's like saying if you can reuse a tool instead of buying a new one each time, you save lots of money.

Investment in Advanced Technology:

Another important thing this launch showed is that the United States is ready to spend money on new and better technologies for military space activities. It's like saying they're investing in new and cool tools to stay ahead in space compared to other countries.

Space Maneuverability Demonstrated

Also, this launch demonstrated that the X-37B spaceplane is really good at doing different moves in space. It showed that this spaceship can do various tricks while it's up there. It's like watching a dancer show off different cool moves during a performance.

So, in simple words, the Falcon Heavy launch carrying the X-37B brought many advancements. It showed that the rocket can lift heavier things, stressed the importance of reusable rockets for saving money, indicated a willingness to invest in new technologies for space, and displayed the X-37B's skill in doing different maneuvers in space. Overall, it's like a big step forward in using better technology for military space missions.

Potential applications of the Falcon Heavy rocket
Launching Heavy Things

The Falcon Heavy rocket is like a super strong delivery truck for space. It can carry really heavy stuff like satellites and space probes up there. It's like sending big packages into space to explore or study things far away from Earth.

Missions to Other Planets

This rocket can also take people on trips! It might carry astronauts on missions to places like the Moon or even

Mars. It's like a spaceship that can take humans on exciting adventures to other worlds.

Big Eye on the Universe

The Falcon Heavy could launch huge telescopes into space. These telescopes would be like super-powered binoculars that can see far-off things in space very clearly. They'd help scientists learn more about the stars and galaxies.

Defensive Space Weapons

Additionally, this rocket might be used to put special tools in space to protect our planet. These tools, like space-based weapons, could defend against attacks from other places in space. It's like having shields in space to keep us safe.

Chapter 8

The X-37B Spaceplane: Controversies and Criticisms

Controversies surrounding the X-37B program

One big issue about the X-37B program is that it's very secretive. This means that not much information about the spaceplane and its missions has been shared with the public. Because it's classified, people don't know a lot about what it does or why. This secrecy has made some people curious and worried about its actual purpose.

Because of this secrecy, many people have guessed about what the X-37B might be doing up in space. Some think it could be used for military purposes, like having weapons or spying tools in space. This has made some folks concerned about making space more about military activities instead of peaceful exploration.

Some experts are worried that because the X-37B can do different moves in space, it might be used to interfere with other satellites. It's like having a car that could drive too close to others and cause problems. This has raised concerns about the spaceplane's abilities and whether they could be used in harmful ways.

Criticisms of the X-37B program

Some experts think the X-37B program is too expensive. It has cost a lot of money, hundreds of millions of dollars, and these experts believe that this money could be better used for other space projects. It's like saying they're spending too much money on this one thing when there are other important things they could spend it on.

Lack of Transparency and Accountability:

Another criticism is that the X-37B program is not very clear about what it's doing or why. There's not much information shared with the public about the spaceplane's missions. People want to know more about how their money is being used, and this lack of information makes them feel like they're not being told everything.

Concerns about Space Weaponization

Some experts are worried that the X-37B might be used as a weapon in space. They're concerned that its capabilities could be used for harmful purposes. It's like having a tool that could potentially cause trouble if not used properly, and this worries them.

In simple words, the criticisms of the X-37B program are that it costs a lot of money, which some think could be spent better elsewhere. There's also criticism about not being clear enough about what the spaceplane does and concerns about its potential use as a weapon in space. Overall, these criticisms revolve around the program's cost, lack of transparency, and worries about its potential uses.

Conclusion

The book delves into the fascinating story of the Falcon Heavy launch carrying the X-37B spaceplane, exploring its design, capabilities, controversies, and potential applications. It covers the technical aspects of both the Falcon Heavy rocket and the X-37B spaceplane, detailing their features, functionalities, and significance in modern space technology. Moreover, it discusses the controversies, criticisms, and speculations surrounding the X-37B program, particularly its secrecy, cost, and potential military applications.

Final Thoughts on the Falcon Heavy Launch of the X-37B Spaceplane:

The Falcon Heavy launch carrying the X-37B marked a significant achievement in space exploration, showcasing the impressive capabilities of the rocket in lifting heavy payloads and demonstrating its potential for various space missions. However, the secrecy surrounding the X-37B program has led to speculation and concerns about its purpose, raising debates regarding transparency, accountability, and the militarization of space. Despite controversies, the launch remains a pivotal moment in advancing space technology and exploration.

Future of the X-37B Program

The future of the X-37B program holds possibilities for further advancements in space technology. There is anticipation regarding the continuation of its missions, potential developments in maneuverability, and its role in future space exploration and defense. However, the program faces challenges in addressing criticisms related to transparency, cost-effectiveness, and the responsible use of its capabilities, which may influence public perception and policy decisions regarding its future endeavors.

In conclusion, the Falcon Heavy launch carrying the X-37B was a significant milestone in space technology, yet the program faces uncertainties concerning transparency, cost, and its role in the evolving landscape of space exploration and defense. The book provides insights into this intriguing intersection of technology, secrecy, and the future of space endeavors.

Frequently asked questions

What is the Falcon Heavy rocket?

The Falcon Heavy rocket is a heavy-lift launch vehicle developed and manufactured by SpaceX. It's composed of three reusable Falcon 9 rocket cores that together generate immense thrust, enabling it to carry heavy payloads into space. It's among the most powerful operational rockets globally and is designed for various missions, including satellite launches and interplanetary missions.

How does the X-37B spaceplane work?

The X-37B spaceplane operates similarly to a miniature version of the Space Shuttle. It's launched atop a rocket, conducts its mission in space for an extended period, and then returns to Earth autonomously. It's highly maneuverable and versatile, equipped with advanced technologies for various experiments, testing, and potential military applications.

Why is military spaceflight important?

Military spaceflight plays a crucial role in national security, surveillance, reconnaissance, communication, and technological advancements. It allows nations to monitor potential threats, protect assets in orbit, gather intelligence, and test advanced technologies critical for defense and strategic purposes.

What are the controversies around X-37B?

The X-37B program is shrouded in secrecy, leading to speculation and controversies regarding its actual missions and capabilities. Critics question its undisclosed objectives, raising concerns about the militarization of space, potential weaponization, and the lack of transparency in its operations.

What are the potential uses of X-37B?

The X-37B's potential applications include conducting experiments in space, testing new technologies, surveillance, reconnaissance, satellite deployment, and potentially serving as a platform for space-based weapons. Its versatility allows it to support various missions in Earth's orbit.

How secretive is the X-37B program?

The X-37B program is highly classified, with limited information publicly available about its specific missions, objectives, or payloads. This secrecy has led to speculation and debate about its actual purposes and activities during its extended missions in space.

What are the advantages of reusable rockets?

Reusable rockets, like the Falcon Heavy, significantly reduce the cost of space missions by allowing the same hardware to be used for multiple launches. This cost-effectiveness enhances access to space, promotes sustainability, and encourages frequent space exploration and utilization.

Are there risks associated with spaceplane missions?

Spaceplane missions, like any space-related endeavor, carry inherent risks. Potential risks include technical malfunctions, launch failures, re-entry issues, and uncertainties in autonomous operations. However, rigorous testing and safety protocols aim to mitigate these risks.

Can Falcon Heavy launch crewed missions?

Yes, the Falcon Heavy has the capability to launch crewed missions. SpaceX has designed the rocket to potentially carry astronauts to destinations such as the Moon or Mars. However, as of now, it hasn't been used for crewed missions but remains an option for future manned spaceflights.

What's the future of military space technology?

The future of military space technology is poised for significant advancements. It will likely witness the integration of more sophisticated satellites, improved space surveillance, enhanced communication systems, the development of anti-satellite capabilities, and further exploration of space for strategic and defense purposes. Collaboration between governments and private entities may also shape the evolution of military space technology.